INNOVATION HINSDALE COLLECTION:

O6 SERIES
PHOTO ARCHIVES

2025
BY KENNETH UPDIKE AND SARAH TOMAC

Thank you to all that help to keep the International Harvester History alive.
Without you, the history would be another by-line in a dusty book. Forgotten.
Sarah Tomac

All serial number and production Data in this book has originated from official IH records.

IBSN 9781952265099
Printed and Distributed by 1831 Press.

This book would not be possible with the help of some retired Hinsdale employees.
To mention them in anything but random order would not be right.
So, thanks to Alan Leupold, Ben Coats, Rich Hale, Mike Thurow, Don Schweiger, Tom McNaull
and Mike Stockwell to mention just a very few.
Thanks!
Kenneth Updike

Contents

Foreward:

This book is the product of many years searching and saving of valuable IH history. To create this would not have been possible without the help from a number of former Hinsdale employees.

The photos themselves are priceless from a historical standpoint.

Photography at IH's Hinsdale Engineering Center was strictly forbidden (for corporate secrecy reasons). Typically when a project was finished the photos were to be destroyed.
This collection of photos were (luckily) saved from that fate, although there are many, many more who were saved but are severely damaged and not presentable.
Having visited the Hinsdale Engineering facility several times in the CaseIH era, there were a few traces of IH still left, if you looked really hard.
The innovative ideas, designs and methods created and tested by the many men and women working there are still being discovered and used yet today.
The format of this book (first in a series of many forthcoming books) is to show the various tractors that IH had designs for. Not all of them made it to the marketplace.
Some existed only on paper or a patent application.
The photos are presented as IH captured them in the time order they were taken. This shows how the designs changed subtly or profoundly over time and during production.
The photo selection is NOT a complete record for each model.
These are the ones that have been saved and shared with me. Many more are missing or still to be found. If/when more surface they will be added in a revised edition.
I hope you enjoy this photo collection of IH tractor development history as much as this author has had finding, collecting and sharing them with you.

Kenneth Updike

Hinsdale Notes

While this photo collection is not complete it does represent some of IH's ideas and progress in their tractor line. Typically the advanced Engineering Group worked on ideas/machines that were at least 10 years into the future. A new tractor development line was usually in this timeframe too.

The time to find/test and adapt new technology was quite long. The ability to get corporate funding to do these projects was a BIG struggle at IH. Many times funding was cut short or stopped and the project died or was delayed.

Going thru project records at Hinsdale, this author sees evidence of new tractor series being started well over 10 years before they came to market. Conversely, some projects were finished thru to completion and then marketing decided to go another way, the market potential had changed OR the funding to see the project thru to production was cut or eliminated.

IH's products may seem dated (at times they were) but funding (or lack of) was usually the culprit behind this.

The Engineers at Hinsdale were creative, caring folks who wanted and designed some of the most innovative farm equipment made. It is truly sad that the lack of corporate funding (in many cases) delayed or stopped designs from becoming reality.

The photos themselves tell the story, to describe the photos as a whole would be tiresome as well as a risk, interpreting the reason for the photo or reducing the intention of the photo, with unneeded words.

606

The 606 was offered as an International (Utility model) only. It replaced the I-460 Utility tractor. The 606 was built from 1961 to 1967. The 606 is rated at 54 PTO HP.

INTERNATIONAL 606
Serial tag begins with "I-606"
Serial Number from 501 to 7939

First I-606 Gas built December 29, 1961 is   Serial No. 501
First I-606 Diesel built April 16, 1962 is   Serial No. 506
First I-606 LPG June 25, 1962 is   Serial No. 922
First I-2606 built December 29, 1961   Serial No. 501
First I-2606 Diesel built June 12, 1962   Serial No. 851

Built from 1961 to 1967, 7,439 total tractors were built.

Page 12

606 Industrial - Model named 2606.

Polyurethane foam baffling between radiator side channels and vertical side sheets of International 606 Diesel, QFE 3401, to block off voids of radiator to prevent re-circulation of hot air into radiator core. Tractor cooling was improved approximately 6F.
Docket 01-000-3607
3118-1-213-July 62

ENGR. CENTER
TEST LAB
HINSDALE, ILL.

Text Reads - Polyurethane foam baffling between radiator side channels and vertical side sheets of International 606 Diesel, QFE 3401, to block off voids of radiator to prevent re-circulation of hot air into radiator core. Tractor cooling was improved approximately 6*F.
Docket 01-000-3607
3118-1-213-July 62.

706

The 706 replaced the model 560 tractor in hp rating.
Built from 1963 to 1967, the 706 was offered as a Farmall (row crop) or International (standard tread) versions. An Industrial version was also made.
The 706 also offered Hi-Clear and FWA (Front wheel assist) versions.
 The 706 is rated at 75 PTO HP.
The 706 was offered in gasoline, diesel and LPG engine power.

Farmall 706 Serial numbers 501 thru 506 were built previously, not on the line.
The first Farmall 706 off the production line is serial 507.

FARMALL 706

Serial tag begins with "F-706"
Serial Number from 501 to 46647

First Farmall 706 built June 3, 1963

First F-706 Gas built June 3, 1963 is	Serial No. 507
First F-706 Diesel built June 3, 1963 is	Serial No. 508
First F-706 LPG built June 3, 1963 is	Serial No. 510
First F-706 Diesel HiClear June 18, 1963 is	Serial No. 652
First F-706 Diesel All Wheel Drive, January 14,1964 is	Serial No. 7583
First F-706 with D-310 Engine, October 28, 1966 is	Serial No. 37237

Built from 1963 to 1967, 46,147 total tractors were built.

Text Reads "1-62"

Text Reads "282 G 1-62"

Text Reads "1-62"

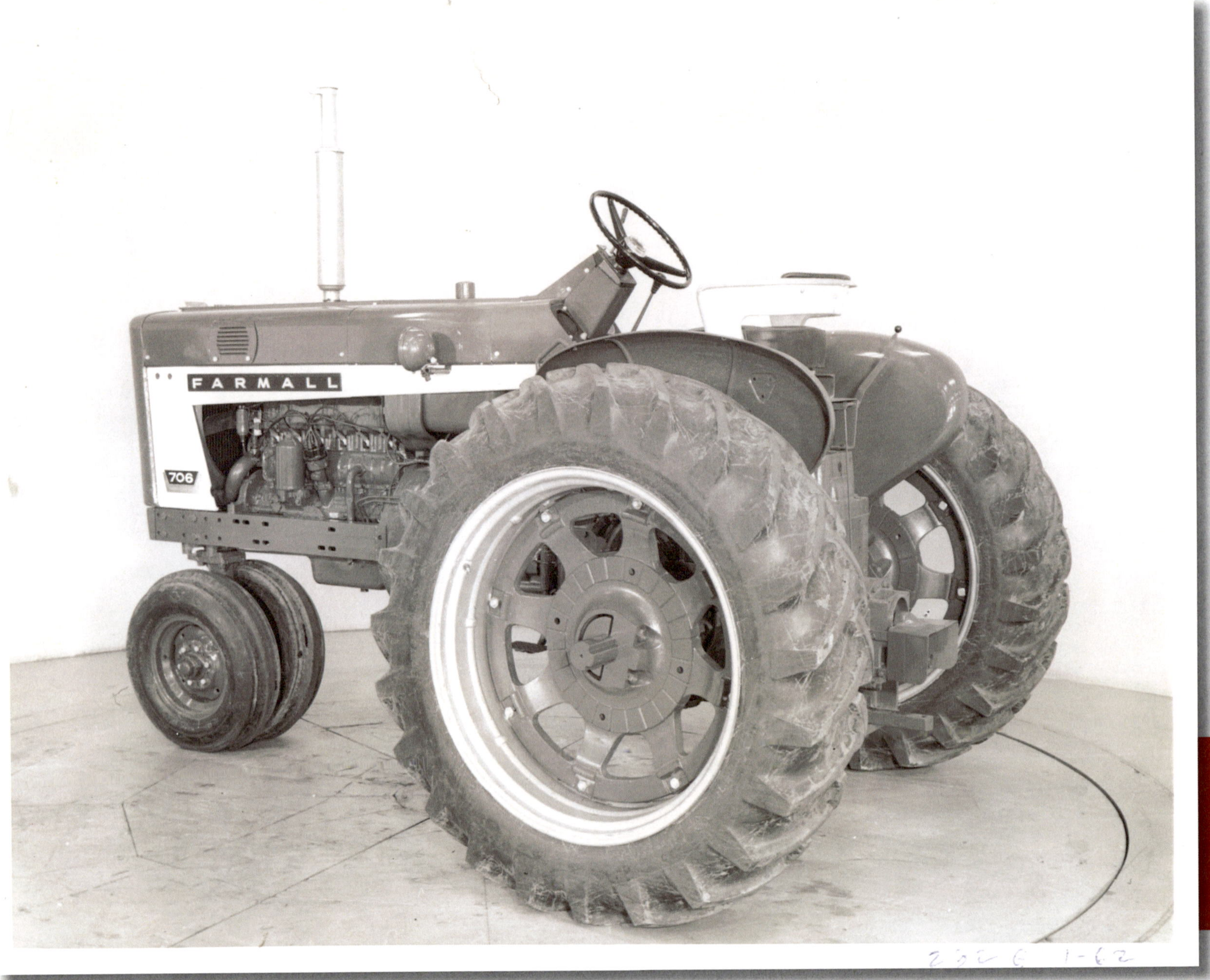

Text Reads "282 G 1-62"

Experimental Farmall 706 Tractor - D-282 Engine0   1-12-62
TT-8-270

Experimental Farmall 706 Tractor - D-282 Engine   1-12-62
                                                    TT-8-271

Experimental Farmall 706 Tractor - D-282 Engine   1-12-62
                                                  TT-8-272

Experimental Farmall 706 Tractor – D-282 Engine   1-12-62
TT-8-273

Text on photograph: FARMALL 706 DIESEL TRACTOR 11-15-62 TP-8-1841

Text Reads: Farmall 706 Diesel Tractor - 11-15-62
TP-8-1841

FARMALL 706 DIESEL TRACTOR

11-15-62
TP8-1851

FARMALL 706 DIESEL TRACTOR

11-15-62
TP8-1852

Text Reads: Farmall 706 Diesel Tractor - 11-15-62
TP8-1852

Text Reads: Farmall 706 Gasoline Tractor -   11-21-62   TP-8-1901

Text Reads: Farmall 706 Gasoline Tractor -   11-21-62   TP-8-1902

FARMALL 706 GASOLINE TRACTOR   11-21-62
                                      TP-8-1903

Text Reads: Farmall 706 Gasoline Tractor -   11-21-62   TP-8-1903

Text inside photo label: FARMALL 706 GASOLINE TRACTOR   11-21-62   TP-8-1904

Text Reads: Farmall 706 Gasoline Tractor -   11-21-62   TP-8-1904   Page 35

FARMALL 706 GASOLINE TRACTOR   11-21-62
TP-8-1905

FARMALL 706 GASOLINE TRACTOR   11-21-62
TP-8-1906

FARMALL 706 GASOLINE TRACTOR 11-21-62 TP-8-1907

FARMALL 706 GASOLINE TRACTOR   11-21-62
TP-8-1908

FARMALL 706 GASOLINE TRACTOR   11-21-62
TP-8-1909

FARMALL 706 GASOLINE TRACTOR   11-21-62
                                      CD-8-1910

FARMALL 706 GASOLINE TRACTOR   11-21-62
TP-8-1911

Page 42

FARMALL 706 GASOLINE TRACTOR   11-21-62
TP-8-1912

FARMALL 706 GASOLINE TRACTOR
11-21-62
TP-8-1913

Text Reads:
Farmall 706 Gasoline Tractor
11-21-62
TP-8-1914

DUAL REAR WHEELS W/DISC OUTER WHEEL   11-19-63
                                                        SU-8-466

DUAL REAR WHEELS W/DISC OUTER WHEEL   11-19-63
SU-8-467

DUAL REAR WHEELS W/DISC OUTER WHEEL

11-19-63
SU-8-468

Photo handwritten label: 8-31-64 TP-8-41/42 HINGED TYPE OIL COOLER ON FARMALL 706 TRACTOR

Text Reads:

Hinged Type oil cooler on
Farmall 706 Tractor

8-31-1964
TP-8-4142

Text Reads:

Hinged Type oil cooler on
Farmall 706 Tractor

8-31-1964
TP-8-4143

International 706

INTERNATIONAL 706

Serial tag begins with "I-706"
Serial Number from 501 to 5,988
First International 706 built June 3, 1963

First I-706 Gas built June 3 1963 is	Serial No. 501
First I-706 Diesel built June 3, 1963 is	Serial No. 502
First I-706 LPG built July 10, 1963 is	Serial No. 588
First I-706 Diesel All Wheel Drive built January 14, 1964 is	Serial No. 1301
First I-706 with D-310 Engine is	Serial No. 5274

Built from 1963 to 1967, 5,488 total tractors were built.

Text Reads:
Int. 2706 Industrial Tractor.
9-27-63
TP-8-3048

INT. 2706 INDUSTRIAL TRACTOR   9-27-63
                                      TP-8-3049

INT. 2706 INDUSTRIAL TRACTOR 9-27-63
TP-8-3050

INT, 2706 INDUSTRIAL TRACTOR   9-27-63
TP-8-3053

INT. 706 TRACTOR W/DUAL REAR WHEELS   10-8-63
TP-8-3107

INT. 706 TRACTOR w/DUAL REAR WHEELS   10-8-62
                                           TP-8-3108

INT. 706 TRACTOR W/DUAL REAR WHEELS   10-8-63
TP-8-3109

INT. 706 TRACTOR W/DUAL REAR WHEELS   10-8-63
TP-8-3110

INT. 706 TRACTOR W/DUAL REAR WHEELS   10-8-63
TP-8-3111

INT. 706 TRACTOR W/DUAL REAR WHEELS   10-8-63
TP-8-3113

INT. 706 TRACTOR W/DUAL REAR WHEELS

10-8-63
TP-8-3114

Text Reads:
Int. 706 tractor w/ dual rear wheels

10-8-63
TP-8-3114

Text Reads:
Int. 706 tractor w/ dual rear wheels

10-8-63
TP-8-3115

INT. 706 TRACTOR W/DUAL REAR WHEELS   10-8-63
TP-8-3116

Text Reads:
I-706D - Shock Mounted Headlamps

2-19-64

TP-8-3783

I-706D - SHOCK MOUNTED HEADLAMPS   2-19-64
TP-8-3784

I-706D - SHOCK MOUNTED HEADLAMPS   2-19-64
TP-8-3785

I-706D - SHOCK MOUNTED HEADLAMPS   2-19-64
                                          TP-8-3786

Hand Written Notes read: Add 1/4 -20 Tapped hole. ; Remove Portion of casting to allow for headlamp clearance; Add 5/16-18 Tapped Hole for Brkt. mounting.; For grille mounting on (706 only)

INTER. 706 DIESEL TRACTOR (NOV. 1, 1964 PRODUCTION)   8-21-64
TP-8-1404

INTER. 706 DIESEL TRACTOR (NOV. 1, 1964 PRODUCTION)   8-21-64
TP-8-1405

Page 78

Text on photograph (handwritten, vertical):
INTER. 706 DIESEL TRACTOR (Nov. 1, 1964 PRODUCTION) 8-21-64 TP-8-1408

Text Reads:
Inter 706 Diesesl Tactor
(Nov 1, 1964 production)
8-21-64
TP-8-1408

INTER. 706 DIESEL TRACTOR (NOV. 1, 1964 PRODUCTION) 8-21-64 TP-8-1409

On image (handwritten, rotated): INTER. 706 DIESEL TRACTOR (NOV. 1, 1964 PRODUCTION) 8-21-64 TP-8-1410

Text Reads:
Inter 706 Diesesl Tactor
(Nov 1, 1964 production)
8-21-64
TP-8-1410

Farmall 706 Hi Clear

806

The 806 was an all new, higher HP tractor.
Built from 1963 to 1967, the 806 was offered as a Farmall (row crop) and International (standard tread) versions. An Industrial version was also made.
The 806 also offered Hi-Clear and FWA (Front Wheel Assist) versions.
The 806 is rated at 95 PTO HP. The 806 was offered in gasoline, diesel and LPG engine power.

The first five Farmall 806 tractors, #501 thru 505, were built previously and not on the line. Records show that the first Farmall 806 off the production line was Serial No.506, a gas tractor, on June 11, 1963.

The first International 806 tractor off the line was Serial No. 502, a diesel tractor, on June 11, 1963. #501 I-806 was built previously, not on the line.

The first Industrial 806, the 2806 Gas, Serial no. 1166 on November 15, 1963.

First 2806 Diesel, Serial No. 803, September 19, 1963.

FARMALL 806

Serial tag begins with "F-806"
Serial Number from 501 to 43,458
First Farmall 806 built June 11, 1963

First 5 Farmall 806 tractors were hand built, Serial numbers. 501-505. Built prior to the Production date.

First F-806 Gas built June 11, 1963 is   Serial No. 506
First F-806 Diesel built June 11, 1963 is   Serial No. 507
First F-806 LPG built August 1, 1963 is   Serial No. 531
First F-806 Gas Hi-Clear, built June 18, 1963   Serial No. 526
First F-806 Diesel All Wheel Drive built January 13, 1964 is Serial No. 5038

Built from 1963 to 1967, 42,958 total F-806 tractors were built.

The 4 Millionth IH Tractor made was a Farmall 806.

F-806 Tractor   9-25-61
TP-8-1138

F-806 Tractor   9-25-61
                       TP-8-1139

F-806 Tractor   9-25-61
TP-8-1140

F-806 Tractor   9-25-61
TP-8-1141

Farmall 806 Tractor - 361 Engine   1-8-62
                                          TT-8-255

Farmall 806 Tractor - 361 Engine   1-8-62
                                          TT-8-256

Farmall 806 Gasoline Tractor
2-5-63
TP-8-2150

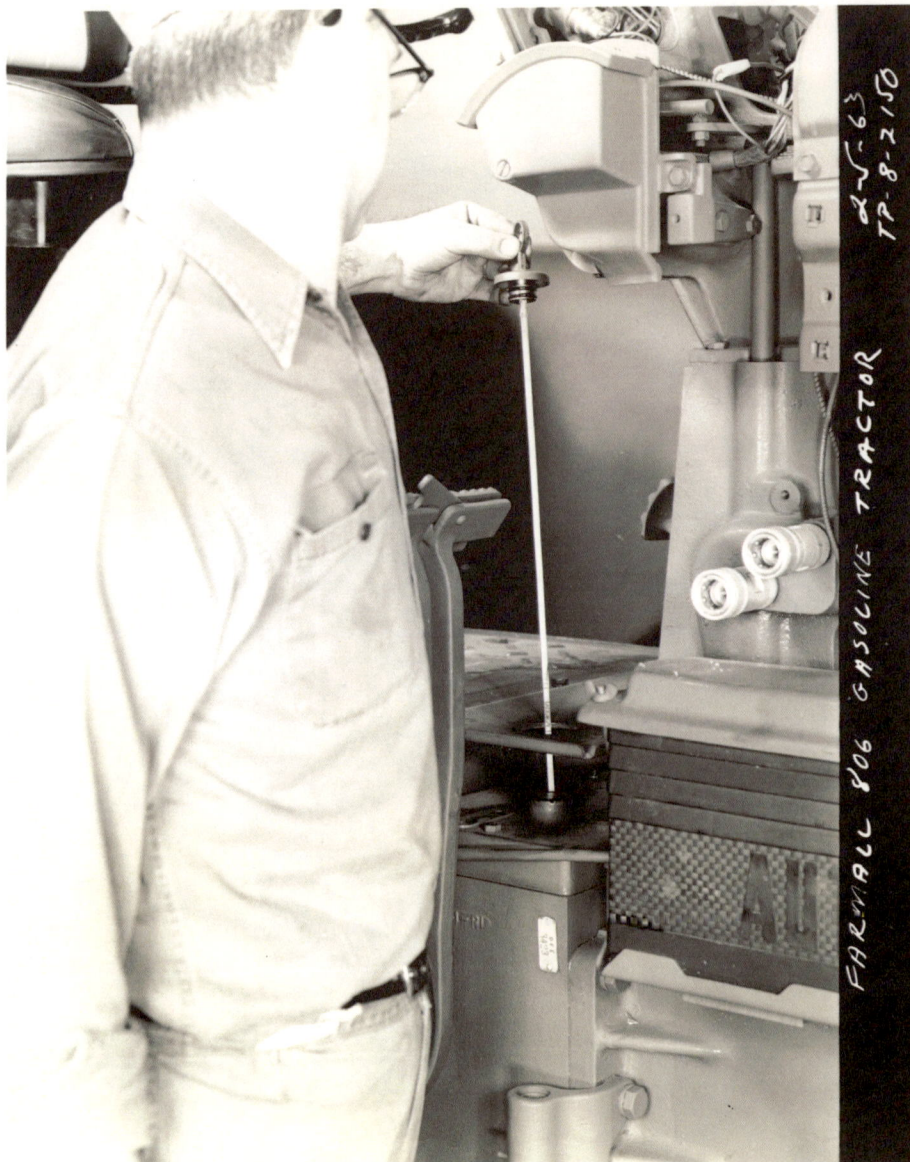

Text on photograph edge: 9-20-63 / TP-8-2968 / INT. 806 ALL WHEEL TRACTOR

Text Reads:
Int. 806 All Wheel Tractor
9-20-63
TP-8-2968

CARMALL 806 TRACTOR W/ADJUSTABLE TREAD WIDE FRONT AXLE   9-27-63   TP-8-3046

FARMALL 806 TRACTOR W/ADJUSTABLE TREAD WIDE FRONT AXLE 9-27-63 TP-8-3047

Text Reads:
Farmall 806 with hand operated park lock
TP-8-3121
10-10-63

Farmall 806 with hand operated park lock

TP-8-3122
10-10-63

Text Reads:
Farmall 806 Gas Tractor
(Nov. 1, 1964 Production)
8-18-64
TP-8-4072

Text on photo: FARMALL 806 GAS TRACTOR (NOV. 1, 1964 PRODUCTION)   8-18-64   TP-8-4073

Text Reads: Farmall 806 Gas Tractor (Nov. 1, 1964 Production)

8-18-64
TP-8-4073

Text Reads: Farmall 806 Gas Tractor (Nov. 1, 1964 Production)

8-18-64
TP-8-4074

Text Reads: Farmall 806 Gas Tractor (Nov. 1, 1964 Production)

8-18-64
TP-8-4075

Text in photograph: FARMALL 806 GAS TRACTOR (NOV. 1, 1964 PRODUCTION) 8-18-64 TP-8-4076

Text Reads: Farmall 806 Gas Tractor (Nov. 1, 1964 Production)

8-18-64
TP-8-4076

Handwritten on photo: FARMALL 806 GAS. TRACTOR (NOV. 1, 1964 PRODUCTION) TP-8-4077 8-18-64

Text Reads:
Farmall 806 Gas Tractor
(Nov. 1, 1964 Production)

8-18-64
TP-8-4077

Text Reads:
Farmall 806 Gas Tractor
(Nov. 1, 1964 Production)
8-18-64
TP-8-4078

FARMALL 806 GAS TRACTOR
NOV. 1, 1964 PRODUCTION

8-18-64
TP-8-4079

FARMALL 806 GAS TRACTOR
(NOV. 1, 1964 PRODUCTION)

8-18-64
TP-8-4080

FARMALL 806 TRACTOR (GAS.)   8-18-64
(NOV. 1, 1964 PRODUCTION)   TP-8-40081

FARMALL 806 GAS TRACTOR
(NOV 1, 1964 PRODUCTION)
8-18-64
TP-8-4082

FARMALL 806 GAS TRACTOR
(NOV. 1, 1964 PRODUCTION)

8-18-64
TP-8-4083

Text Reads:
Farmall 806 Gas Tractor
(Nov. 1, 1964 Production)
8-18-64
TP-8-4083

Text Reads:
Farmall 806 Gas Tractor
(Nov. 1, 1964 Production)
8-18-64
TP-8-4084

FARMALL 806 DIESEL (NOV. 1, 1964 PRODUCTION)   8-25-64
TP-8-4130

Text Reads:
Farmall 806 Diesel Tractor
(Nov. 1, 1964 Production)
8-25-64
TP-8-4132

Text Reads: Farmall 806 Diesel (Nov. 1, 1964 Production) 8-25-64
TP-8-4138

International 806

Serial tag begins with "I-806"
Serial Number from 501 to 8,553
First International 806 built June 11, 1963

First I-806 Gas built June 14, 1963 is   Serial No. 506
First I-806 Diesel built June 11, 1963 is   Serial No. 502 (#501 is diesel, hand-built,
                                                   not on line)

First I-806 LPG built June 13, 1963 is   Serial No. 504
First I-806 Diesel All Wheel Drive January 13, 1964 is Serial No. 1459

Built from 1963 to 1967, 8,053 total I-806 tractors were built.

Text Reads: GM Diesel 1-62

Text Reads: GM Diesel 1-62

I-806 LP TRACTOR

1-28-64

TT-8-459

I-806 LP TRACTOR   1-28-64
                          TT-8-457

806 All Wheel Drive

Offered both in Farmall 806 and International 806

INT. 806 ALL WHEEL TRACTOR   9-20-63
TP-8-2957

INT. 806 ALL WHEEL TRACTOR   9-20-63
TP-8-2958

INT. 806 ALL WHEEL TRACTOR   9-20-63
TP-8-2960

INT. 806 ALL WHEEL TRACTOR

9-20-63
TP-8-2961

INT. 806 ALL WHEEL TRACTOR   9-20-63
TP-8-2963

INT. 806 ALL WHEEL TRACTOR 9-20-63
TP-8-2965

INT. 806 ALL WHEEL TRACTOR 9-20-63 TP-8-2966

Handwritten on photo edge: 9-20-63 TP-8-2967 / INT. 806 ALL WHEEL TRACTOR

Text Reads:
Int. 806 All Wheel Tractor
                    9-20-63
                    TP-8-2967

INT. 806 ALL WHEEL TRACTOR 9-20-63
TP-8-2971

INTERNATIONAL 806 ALL WHEEL TRACTOR FRONT AXLE   TP-8-3124
10-11-63

INTERNATIONAL 80G ALL WHEEL TRACTOR FRONT AXLE   TP-8-3125
10-11-63

Farmall 806 Turbo

The following is a variety of development photos of the Farmall 806 Turbo Diesel

Text Reads:
Farmall 806 Diesel Tractor
2-2-63
TP-8-2031

Hand written note:
Orentation IPTO Linkage

Text that appears on the photograph: FARMALL 806 DIESEL (NOV. 1, 1964 PRODUCTION) 8-25-64 TP-8-4134

Text Reads:
Farmall 806 Diesel
(Nov. 1, 1964 production)

8-25-64
TP-8-4134

Text Reads:
Farmall 806 Diesel (Nov. 1, 1964 production)

8-25-64
TP-8-4137

Text Reads:
Farmall 806 Diesel (Nov. 1, 1964 production)

8-25-64
TP-8-4136

Text Reads:
F-806D Turbo-Charged
11-20-63
TP-8-3234

F-806 D   TURBO-CHARGED

11-20-63

TP-8-3235

F-806D TURBO-CHARGED

11-20-63
TP-8-3236

F-806D   TURBO-CHARGED   11-20-63   TP-8-3257

F-806D TURBO-CHARGED
11-20-63
TP-8-3238

Identity Tag: QFE 3827
Serial Tag number unreadable.

Text Reads:
F-806D Turbo-Charged
        12-12-63
        TP-8-3245

Text Reads:
F-806D Turbo-Charged
12-12-63
TP-8-3246

2806

An Industrial tractor, the 2806 is modeled from an 806 tractor.

INT. 2806 (DIESEL) INDUSTRIAL TRACTOR   10-2-63
                                               TP-8-3078

Text Reads:
Int. 2806 (Diesel) Industrial Tractor.
10-2-63
TP-8-3079

INT. 2806 (DIESEL) INDUSTRIAL TRACTOR

10-2-63

TP-8-3080

Text Reads: Int. 2806 (Diesel) Industrial Tractor.   10-2-63
TP-8-3081

1206

The need for "more power" lead to the 1206 development.

Initially built as a modified 806 with turbocharger, the added power lead IH to develop a more robust driveline and introduce 38 inch diameter drive wheels.

The 1206 was made from 1965 to 1967 and features a unique red/white paint scheme.

It was offered as a Farmall (row crop) and International (standard tread) versions.

An Industrial version was also made.

The 1206 also had a FWA (Front Wheel Assist) version.

The 1206 is rated at 115 PTO HP.

The 1206 was availible only in diesel engine power.

It was the first turbocharged row crop tractor over the 100 HP level for all tractor makes.

FARMALL 1206

Serial tag begins with "F-1206"
Serial Number from 7501 to 15,903
Diesel fuel only, rated at 115 horsepower.

Pilot model #501 built July 9, 1965.

Farmall 1206 production began September 13, 1965.   Serial No. 7501

Built from 1965 to 1967, 8,403 total tractors were built.

INTERNATIONAL 1206

Serial tag begins with "I-1206"
Serial Number from 7501 to 9090
Diesel fuel only, rated at 115 horsepower.

Pilot model #501 built July 9, 1965,

First International 1206 built September 13, 1965   Serial No. 7501.

Built from 1965 to 1967, 1,590 total tractors were built.

INT. 1206 TURBO TRACTOR

6-17-64

TP-8-3954

Page 192

INT. 1206 TURBO TRACTOR

6-17-64

TP-8-3955

INT. 1206 TURBO TRACTOR

6-17-64
TP-8 - 3956

INT, 1206 TURBO TRACTOR

6-17-64
TP-8-3957

INT. 1206 TURBO TRACTOR   6-17-64   TP-8-3958

INT. 1206 TURBO TRACTOR

6-17-64
TP-8-3960

INT. 1206 TURBO TRACTOR 6-17-64 TP-8-3961

Page 200

Text Reads:
Farmall 1206 Turbo Diesel
1-18-65
TP-8-4475

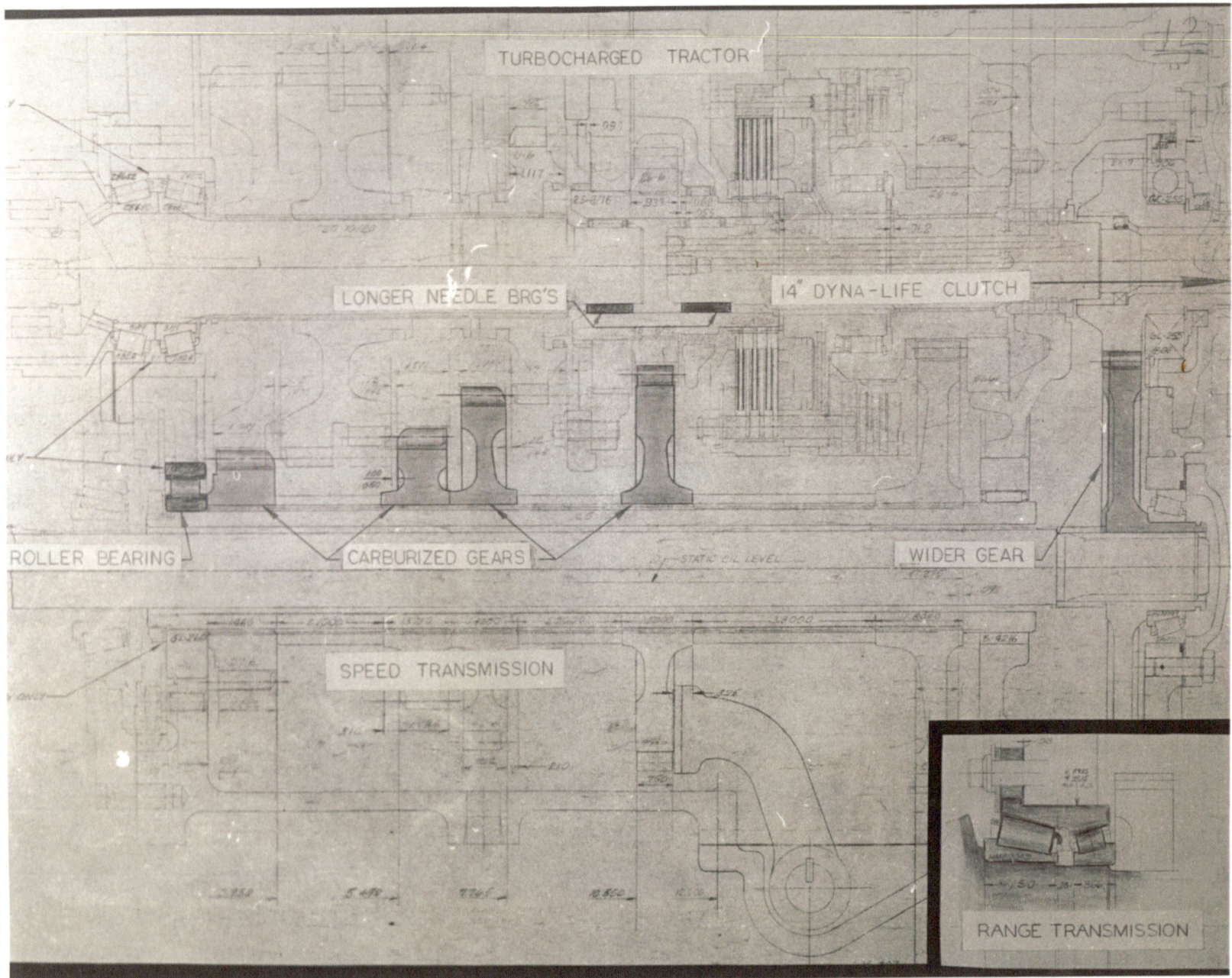

TURBOCHARGED TRACTOR

LONGER NEEDLE BRG'S

14" DYNA-LIFE CLUTCH

ROLLER BEARING

CARBURIZED GEARS

STATIC OIL LEVEL

WIDER GEAR

SPEED TRANSMISSION

RANGE TRANSMISSION

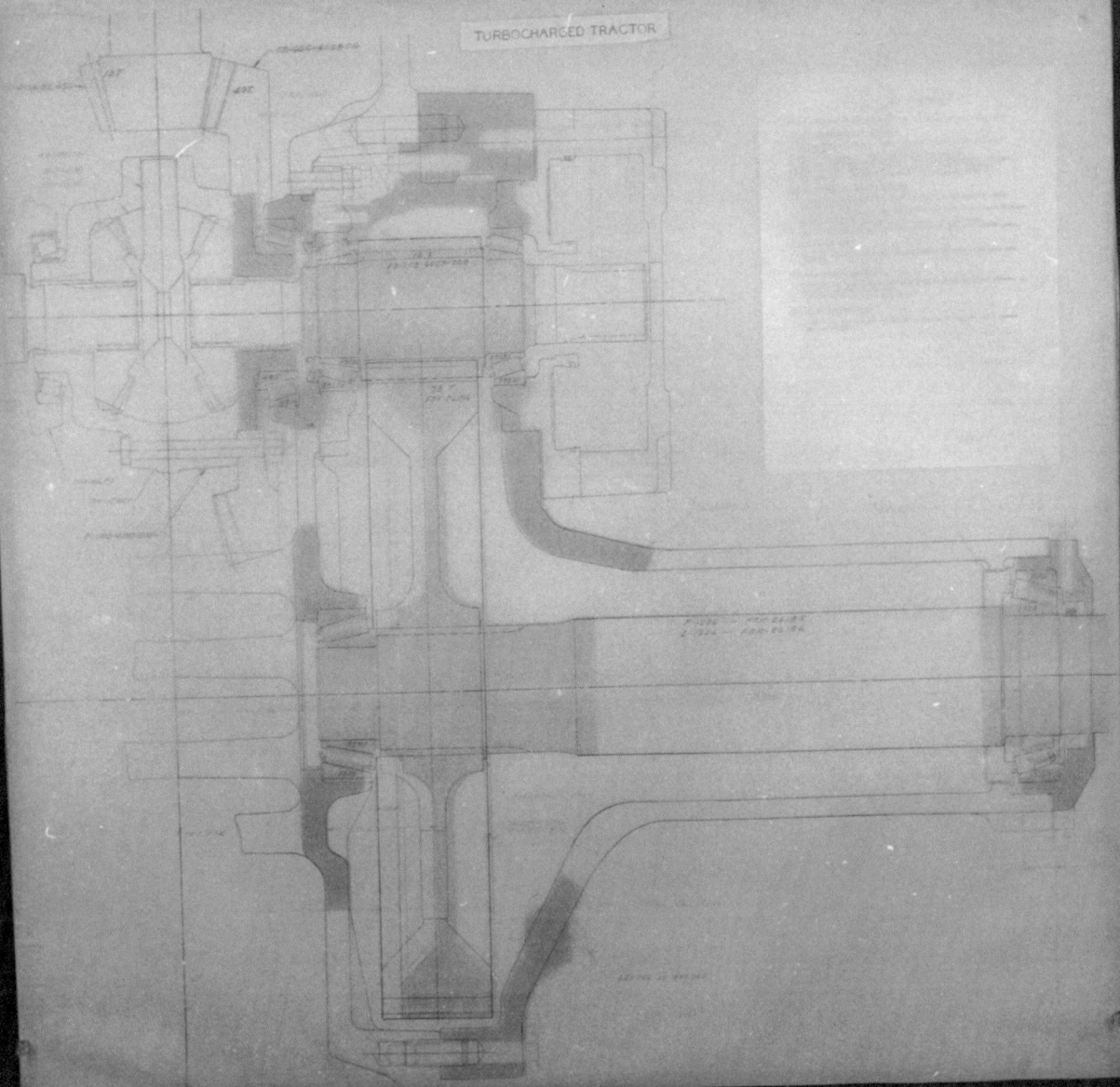
TURBOCHARGED TRACTOR

1206 TURBO

CLUTCH-
14" DIA. DYNA-LIFE

TRANSMISSION - 4 SPEED -
CARBURIZED GEARS
PTO DRIVEN GEAR - WIDER
HEAVIER BEARINGS

FINAL DRIVE -
WIDER BULL GEAR
$^{12}/_{49}$ MAIN SHAFT & DIFF. GEARS
HEAVIER BEARINGS

STYLING-
NEW FRONT HOOD
    ''   GRILLE
    ''   SIDE PANEL
    ''   NAME PLATE (1206 TURBO)

DELUXE FENDER - "F" - OPTIONAL
W/SHOCK PROOF HEAD LAMPS

1206 TURBO

HITCH-

QUICK ATTACHABLE 3 POINT
CATEGORY III CONVERTABLE TO
CATEGORY II.
STORAWAY SWINGING DRAWBAR.
HYDRAULIC PRESSURE — 2000 P.S.I.
HEAVY DUTY VERTICAL ADJUSTABLE
DRAWBAR.
HEAVY DUTY SWINGING DRAWBAR.

I.P.T.O.-

1000 R.P.M.
HYDRAULIC PRESSURE — 200 P.S.I.
NEW I.P.T.O. COVER, SHAFT & SHIELD.

WHEELS-

NEW HEAVIER SECTION REAR
WHEELS.
NEW REAR TIRES 24.5-32
18.4-38
16.9-38

1206 TURBO TRACTOR   2-18-64
TP-8-5365

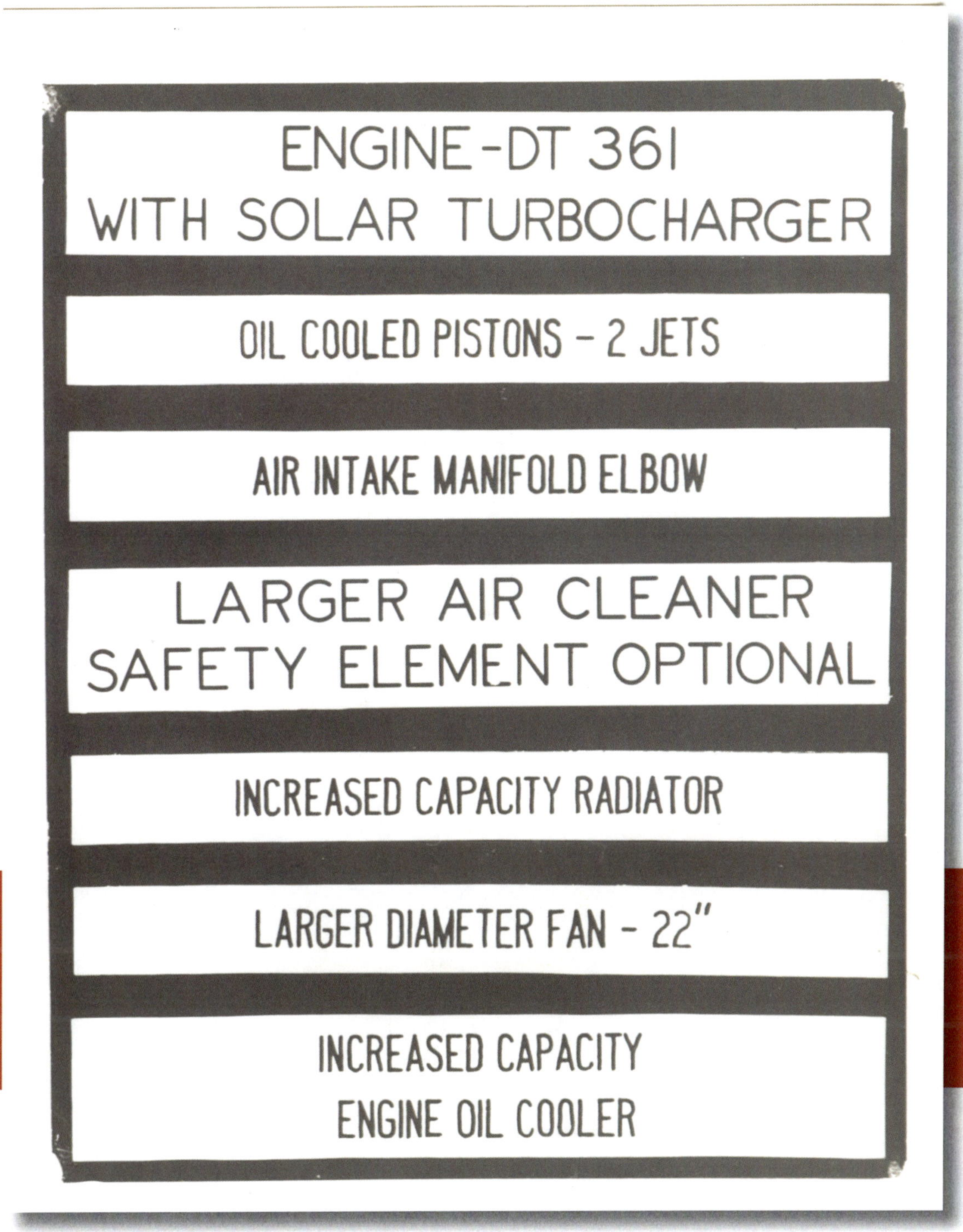

ENGINE-DT 361 WITH SOLAR TURBOCHARGER

OIL COOLED PISTONS – 2 JETS

AIR INTAKE MANIFOLD ELBOW

LARGER AIR CLEANER SAFETY ELEMENT OPTIONAL

INCREASED CAPACITY RADIATOR

LARGER DIAMETER FAN – 22"

INCREASED CAPACITY ENGINE OIL COOLER

1206 TURBO TRACTOR 2-18-64 TP-8-3367

1206 TURBO TRACTOR   2-18-64
TP-8-3368

1206 TURBO TRACTOR   2-18-64
                            TP-8-3369

1206 TURBO TRACTOR 2-18-64 TP-8-3370

Text Reads:
1206Turbo Tractor
2-18-64
TP-8-3371

Text Reads:
1206Turbo Tractor
2-18-64
TP-8-3372

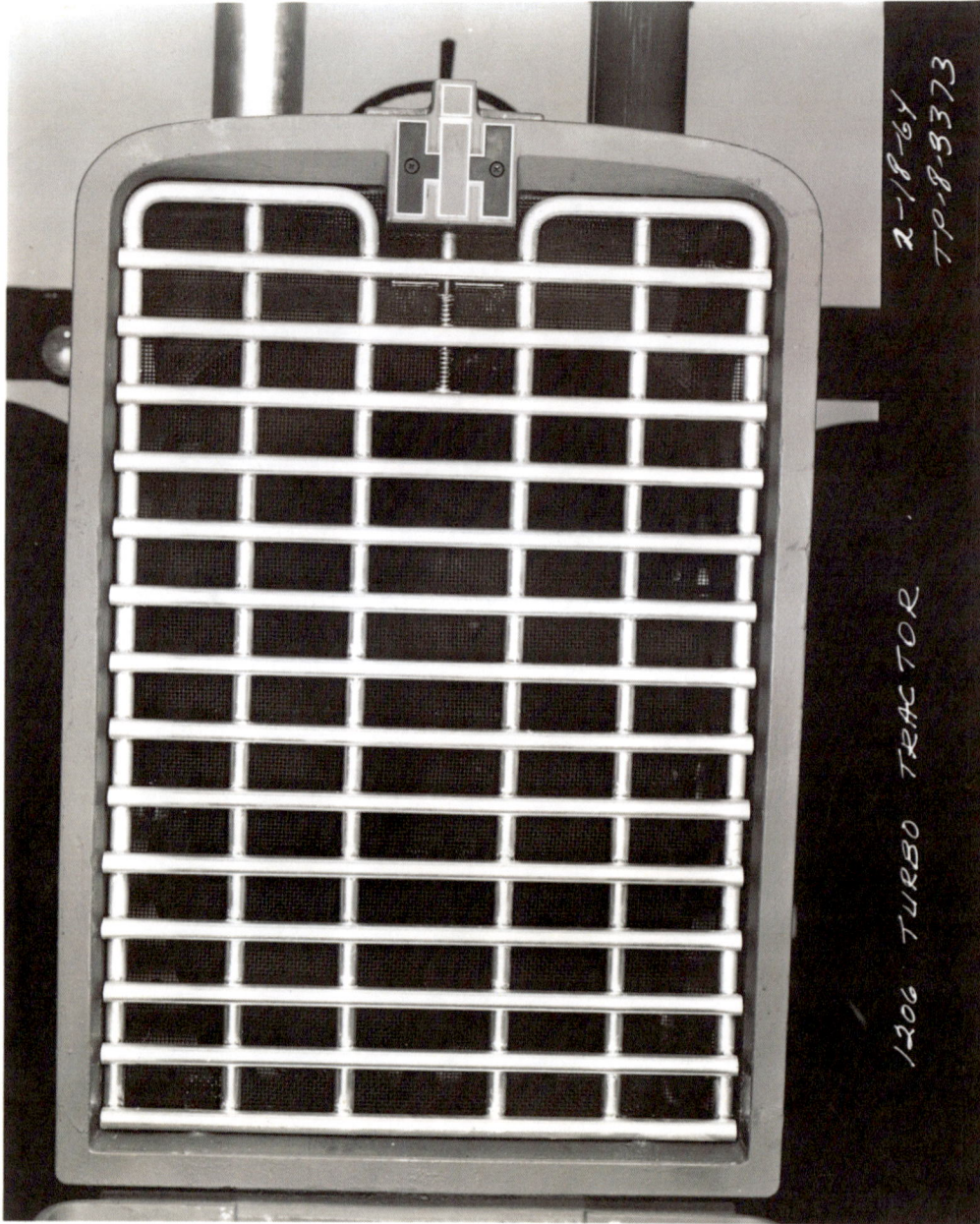

Text Reads:
1206Turbo Tractor
2-18-64
TP-8-3373

Text Reads:
1206 Turbo Tractor
2-18-64
TP-8-3374

1206 TURBO TRACTOR   2-18-64
TP-8-3375

FARMALL 1206 TURBO TRACTOR W/WIDE FRONT AXLE   6-15-64
TP-8-3946

FARMALL 1206 TURBO TRACTOR w/WIDE FRONT AXLE 6-15-64 TP-8-3947

FARMALL 1206 TURBO TRACTOR W/WIDE FRONT AXLE 6-15-64
TP-8-3948

FARMALL 1206 TURBO TRACTOR W/WIDE FRONT AXLE   6-15-64
TP-8-3949

FARMALL 1206 TURBO TRACTOR W/WIDE FRONT AXLE   6-15-64
TP-8-3950

FARMALL 1206 TURBO TRACTOR W/WIDE FRONT AXLE   6-15-64
TP-8-3951

FARMALL 1206 TURBO TRACTOR W/WIDE FRONT AXLE   6-15-64
TP-8-3952

FARMALL 1206 TURBO TRACTOR W/WIDE FRONT AXLE   6-15-64
                                                       TP-8-3953

Text Reads:
INT. 1206Turbo Tractor
6-17-64
TP-8-3959

Bottom Margin:
8-28-65
TP-8-4131

FARMALL 1206 TURBO TRACTOR   6-18-64

TP-8-3963

FARMALL 1206 TURBO TRACTOR 6-18-64 TP-8 3964

FARMALL 1206 TURBO TRACTOR 6-18-64 TP-8-3965

FARMALL 1206 TURBO TRACTOR 6-18-64 TP-9-3966

FARMALL 1206 TURBO TRACTOR

6-18-64

TP-8-3967

FARMALL 1206 TURBO TRACTOR 6-18-64
TP-8-3968

FARMALL 1206 TURBO TRACTOR   6-18-64
TP-8-3969

FARMALL 1206 TURBO TRACTOR   6-18-64
TP-8-3970

F-1206 TURBO TRACTOR

TP-8-4295
10-27-64

F-1206 TURBO TRACTOR

TP-8-4296
10-27-64

Text Reads:
1206Turbo Tractor
TP-8-4297
10-27-64

1206 TURBO TRACTOR

11-6-64
TP-8-4311

1206 TURBO TRACTOR   11-6-64
TP-8-4312

Text Reads:
1206 Solar Turbo Engine
12-3-64
TP-8-4374

Text Reads:
1206 Solar Turbo Engine
12-3-64
TP-8-4375

1206 SOLAR TURBO ENGINE 12-3-64
TP-8-4376

1206 SOLAR TURBO ENGINE   12-3-64
TP-8-437J

Handwritten on photo: 12-3-64 / TP-8-4378 / 1206 SOLAR TURBO ENGINE

Text Reads: 1206 Solar Turbo Engine   12-3-64   TP-8-4374
Serial Tag: F-806. QFE tag and serial number unreadable.

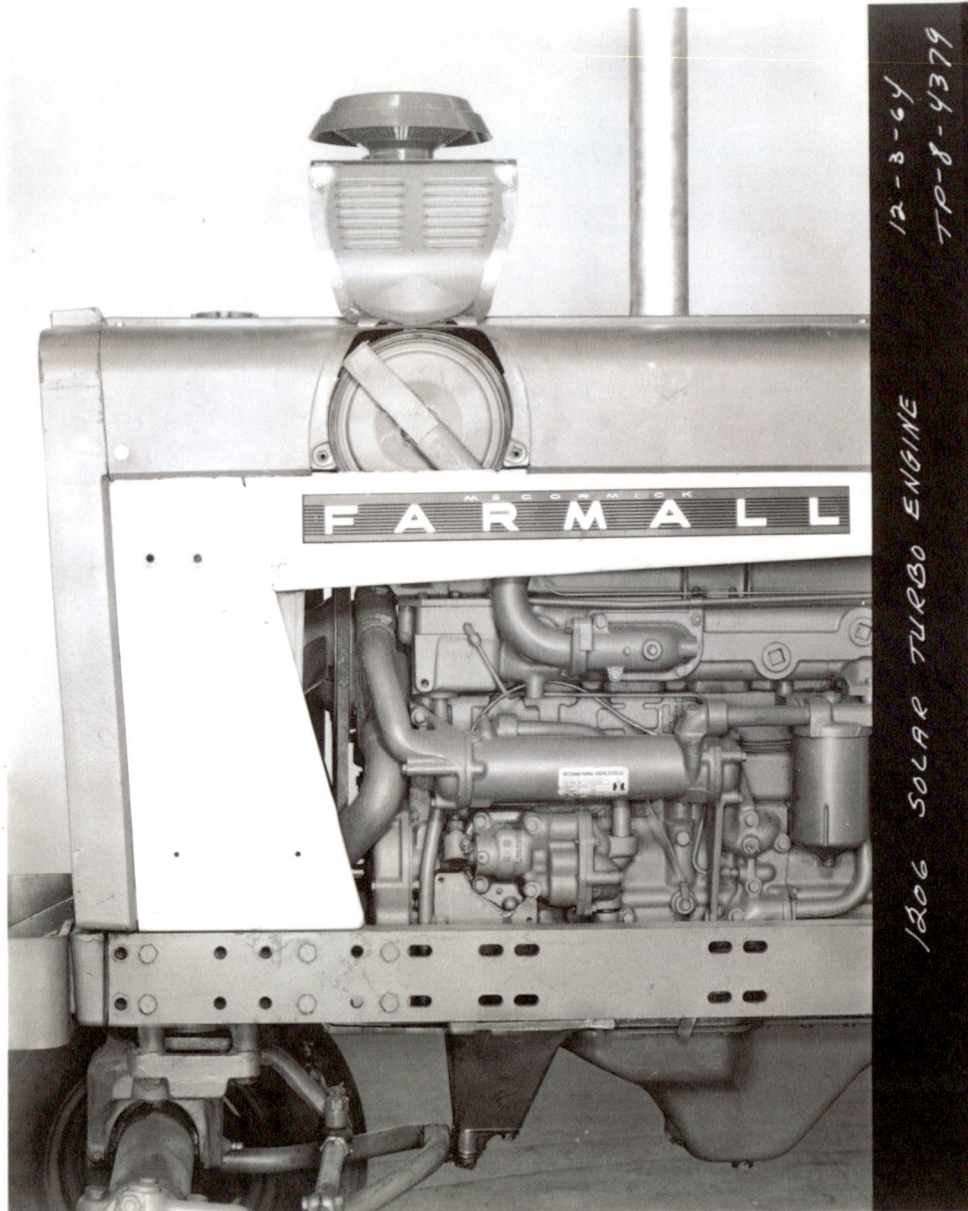

Handwritten on image: 12-3-64 TP-8-4379

Handwritten on image: 1206 SOLAR TURBO ENGINE

M°CORMICK FARMALL

Text Reads:
1206 Solar Turbo Engine
12-3-64
TP-8-4379

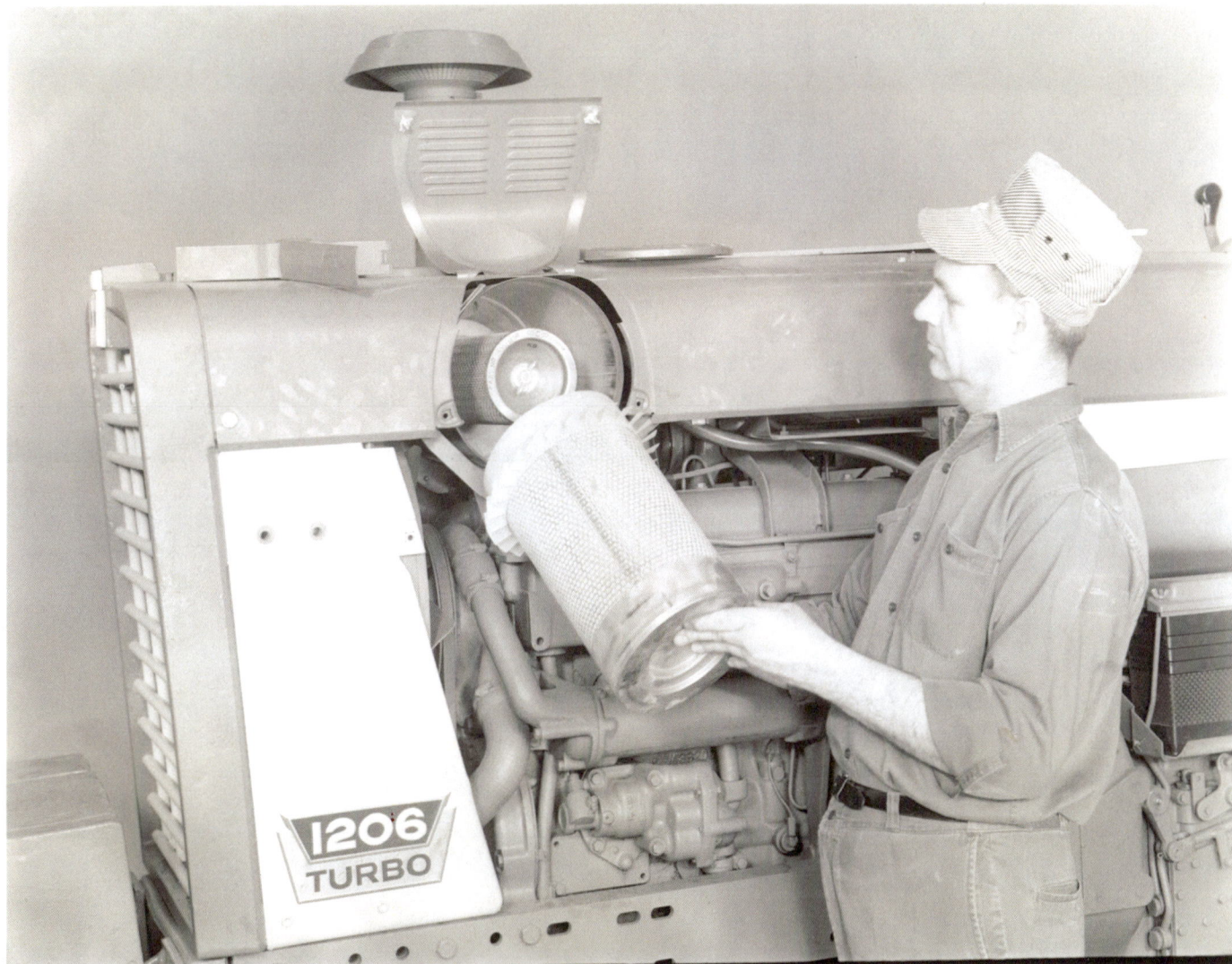

FARMALL 1206 TURBO DIESEL   1-18-65
TP-8-4481

FARMALL 1206 TURBO DIESEL
1-18-65
TP-8-4484

Text Reads: 1206 Solar Turbo Engine 1-18-65 TP-8-4484
Serial Tag: F-806. QFE tag and serial number unreadable.

FARMALL 1206 TURBO DIESEL 1-18-65
TP-8-4-86

FARMALL 1206 TURBO DIESEL 1-18-65
TP-8-4487

FARMALL 1206 TURBO DIESEL   L-18-65
TP-8-4488

FARMALL 1206 TARBO DIESEL 1-18-65 TP-8-4489

FARMALL 1206 TURBO DIESEL

1-18-65
TP-8-4491

Text Reads:
Farmall 1206 Turbo Diesel

1-18-65
TP-8-4499

FARMALL 1206 TURBO DIESEL   1-18-65
TP-8-4494

FARMALL 1206 TURBO DIESEL   1-18-65
TP-8-4495

Text Reads:
Farmall 1206 Turbo Diesel
1-18-65
TP-8-4497

FARMALL 1206 TURBO DIESEL   1-18-65   TP-8-4496

Text Reads:
Farmall 1206 Turbo Diesel
1-18-65
TP-8-4497

Text Reads:
Farmall 1206 Turbo Diesel
1-18-65
TP-8-4498

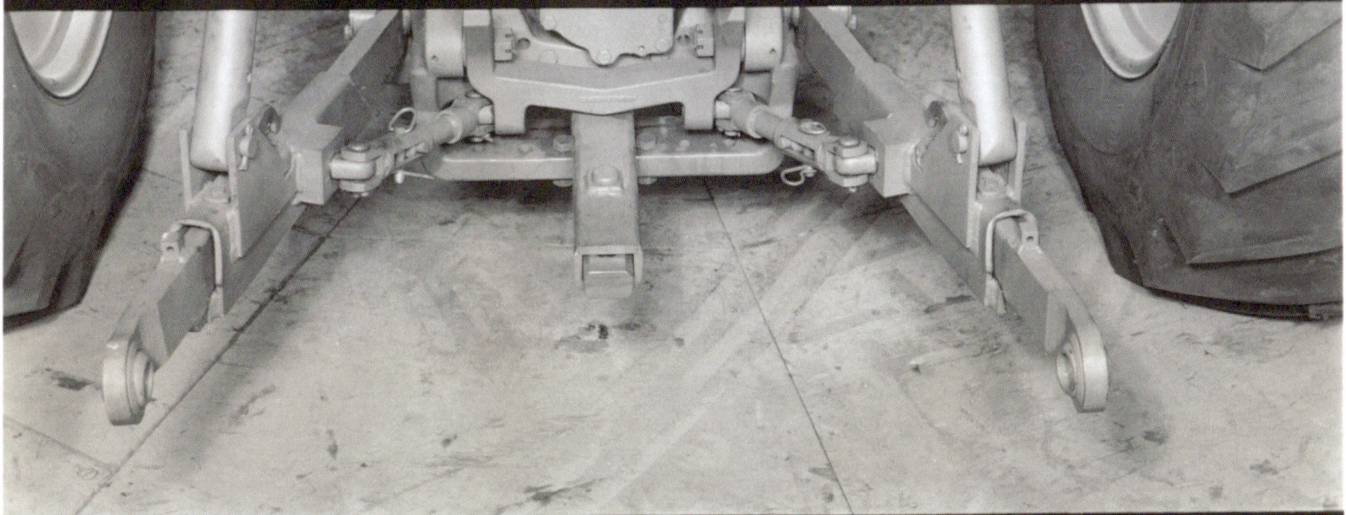

Text Reads: Farmall 1206 Turbo Diesel

1-18-65
TP-8-4502

Text Reads:
Farmall 1206 Turbo Diesel
1-18-65
TP-8-4501

FARMALL 1206 TURBO DIESEL   1-18-65

Text Reads: Farmall 1206 Turbo Diesel   1-18-65

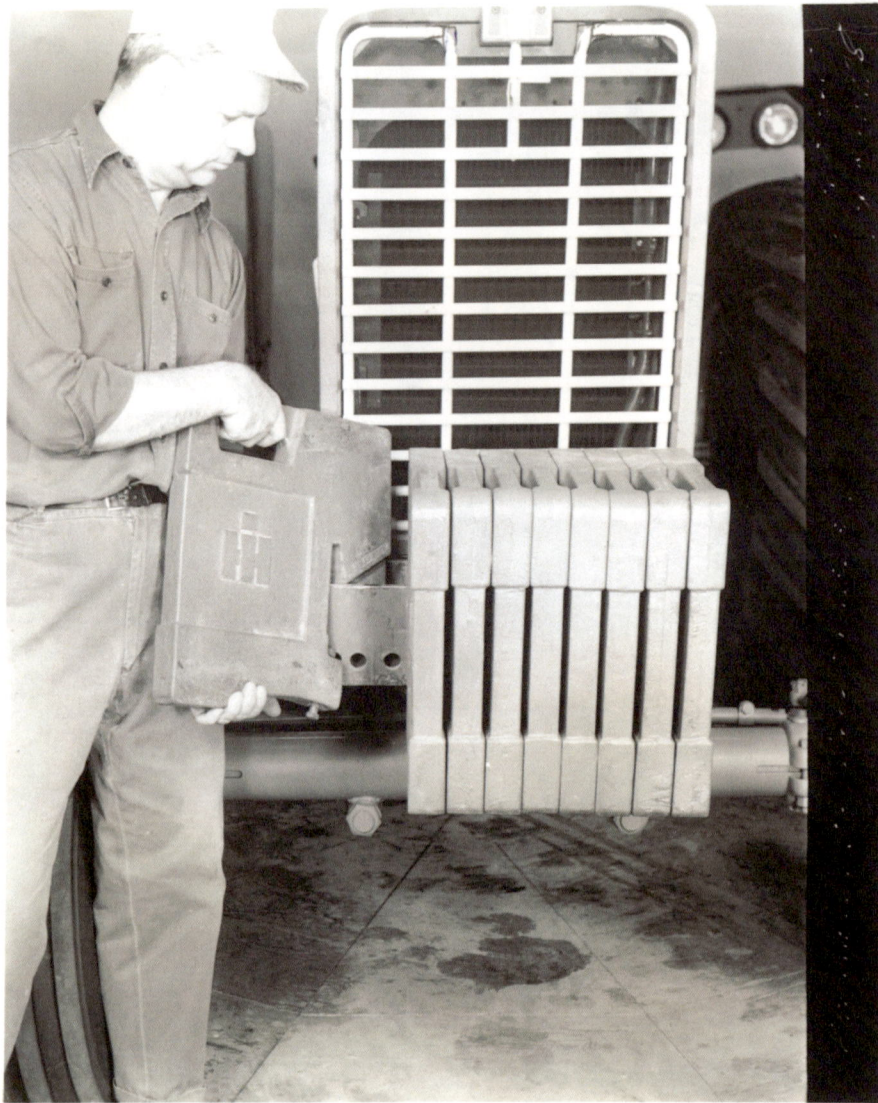

Text Reads:
Farmall 1206 Turbo Diesel
1-18-65
TP-8-4478

F-1206 W/SOLAR TURBO   5-5-65
                              TP-8-4677

F-1206   W/SOLAR TURBO

5-5-65
TP-8-4678

F-1206 W/SOLAR TURBO

6-5-65
TP-8-4679

F-1206 W/SOLAR TURBO   5-5-65
TP-8-4680

F-1206 w/solar Turbo   5-5-65   TP-8-4681
Tag Reads: QFE 3827   Serial Tag: F806D number not visible.

F-1206 W/SOLAR TURBO   5-5-65
TP-8-4682

F-1206 W/SOLAR TURBO   5-5-65
TP-8-4683

F-1206 W/SOLAR TURBO

5-5-65
TP-8-4684

F-1206 w/ SOLAR TURBO

5-5-65
TP-8-4685

F-1206 W/SOLAR TURBO   5-5-65
T-8-4686

F-1206 W/SOLAR TURBO
5-5-65
TP-8-4687

F-1206 W/SOLAR TURBO   5-5-65
TP-8-4688

www.ingramcontent.com/pod-product-compliance
Lightning Source LLC
Chambersburg PA
CBRC100225240326
41458CB00131B/6522